GEORGES LEMAÎTRE

The Big Bang Theory
and the Origins of Our Universe

Written by Pauline Landa
Translated by Jessica Foster

History **50MINUTES.com**

50MINUTES.com

BECOME AN EXPERT
IN HISTORY

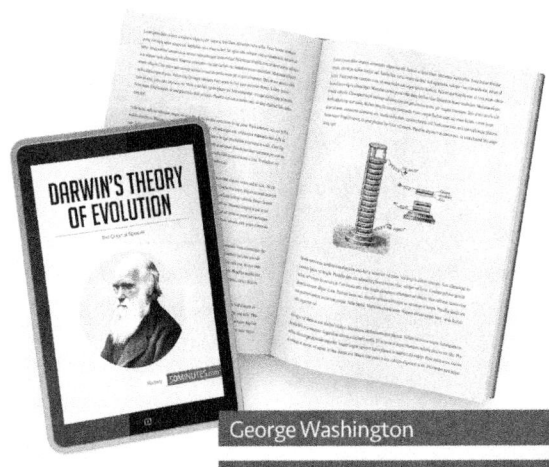

DARWIN'S THEORY OF EVOLUTION

George Washington

The Battle of Austerlitz

Neil Armstrong

The Six-Day War

The Fall of Constantinople

www.50minutes.com

GEORGES LEMAÎTRE AND THE BIG BANG THEORY   1

Key information
Introduction

CONTEXT   3

A unique model for the universe?
The universe was not created in seven days

LEMAÎTRE'S LIFE   7

A keen interest in science
An intellectual and spiritual path
Formative journeys
One man, two callings

LEMAÎTRE'S THEORIES   12

Einstein and Friedmann's contributions
The theory of the expansion of the universe (1927)
The hypothesis of the primeval atom (1931)
The rest of his work
The chronology of the Big Bang

IMPACT   23

Reception among the scientific community: between continuity and criticism
Reception of Lemaître's theories across the world
What remains of the Big Bang theory?

SUMMARY   27

FURTHER READING

GEORGES LEMAÎTRE AND THE BIG BANG THEORY

KEY INFORMATION

- **Born:** 17 July 1894 in Charleroi, Belgium.
- **Died:** 20 June 1966 in Leuven, Belgium.
- **Main achievements:** various theories that enabled advances in research on the universe and its origins, such as those on the expansion of the universe (1927) and the primeval atom (1931).
- **Impact of his research:** the Big Bang theory is now widely accepted and has led to the creation of a new field of research: modern cosmology.

INTRODUCTION

Man has always sought to understand the world and, by extension, the universe surrounding it. Throughout the centuries, various theories have emerged depending on the customs of each period and have given rise to conceptions of the universe that were not always in line with religious and political authorities. Nobody would have thought in the time of Galileo (Italian astronomer and physicist, 1564-1642), condemned by the Church for his writings on heliocentrism, that it would be a Belgian priest, Georges Lemaître, who would be responsible for the Big Bang theory, which is still widely accepted.

His 'hypothesis of the primeval atom', put forward in 1931, dates the origin of the universe to 13.7 billion years ago.

Lemaître theorised that the universe was then compressed into a single atom – a part of the theory that has now been refuted – that a cosmic shock enabled to break up into a large number of electrons, photons, etc., thus giving rise to the universe as we know it today. Lemaître was a pioneer in this subject, especially given that the other scientists of his time were convinced that the universe had always existed and that it was therefore pointless to try to find its beginnings. Following the publication of his research, however, they were all forced to reconsider their positions and accept this new theory that would profoundly influence cosmology and physics in the 20th and 21st centuries.

CONTEXT

A UNIQUE MODEL FOR THE UNIVERSE?

It is impossible to talk about Georges Lemaître without first mentioning one of the greatest scientists of the 20th century, Albert Einstein (1879-1955). It was notably thanks to his theory of general relativity (1915) that Lemaître would go on to develop his hypothesis of the primeval atom. Before the 'everything is relative' period, scientists had been dedicated since Ancient times to trying to understand the world around them by suggesting a model of the universe capable of explaining it.

With Aristotle (Greek philosopher, 384-322 BC) and Ptolemy (Greek astronomer, circa AD 100-170), the prevailing model was the geocentric model, according to which the Earth was situated at the centre of the universe, and this would last until the 16th century. At this time, the model was called into question by academics such as Giordano Bruno (Italian philosopher, 1548-1600) and Galileo in favour of heliocentrism, which led to confrontation with the religious authorities. The former was burned at the stake, the latter condemned by the Church. Their ideas, however, circulated around the scientific community and regularly provided food for debate.

Portrait of Galileo by the painter Justus Sustermans.

Eventually it was Albert Einstein, almost three centuries later, who freed researchers from this search for a unique model as, to him, there was no need to favour one model above another. In fact, each model was based on a coherent frame of reference and it was therefore unnecessary to choose just one, as it would be no more or less valid than another. In his opinion, no one unique model could possibly encompass the whole universe.

In 1927, Lemaître, who was very interested in these works on relativity, developed Einstein's calculations in his article, "A homogeneous universe of constant mass and growing radius accounting for the radial velocity of extragalactic nebulae", and described a universe that is expanding and whose density of matter tends towards minus infinity as we look further back in time. However, his theory of the expanding universe was not particularly well received by the scientists of the time, and even Einstein described it as appalling. It was not until 1929, with the American astronomer Edwin Hubble (1889-1953) and his eponymous Hubble's Law, that this idea of expansion would be accepted by the whole scientific community.

THE UNIVERSE WAS NOT CREATED IN SEVEN DAYS

While today we appreciate the point of dating the origin of the universe, the same was not true of Lemaître's time. In fact, at the beginning of the 20th century, it was still forbidden to mention, in cosmology, metaphysical notions that were normally dictated by religion, such as entirety, time, the beginning, etc. The great cosmologists of the time, including Einstein, refused to believe that there was a precise moment at which the universe appeared. Lemaître was therefore the first to mention the idea of the origin of the universe after he read an article by Arthur Eddington (British astronomer and physicist, 1882-1944) on the end of the world.

Starting from the assumption that there is a constant energy that is distributed in quanta (minimal quantities of energy) throughout the universe and that the number of quanta is always increasing, if we consider the history of the universe, we must be able to trace back an ever-lower number of quanta up to the point at which we arrive at a unique quantum in which the entire universe was concentrated. This is how the hypothesis of the primeval atom was reached in 1931, a theory that is still accepted and now known as the Big Bang theory.

> **DID YOU KNOW?**
>
> Georges Lemaître did not invent the term 'Big Bang', which has stuck. It was in fact one of his critics, the British astronomer Fred Hoyle (1915-2001), who coined this expression. During an interview on BBC radio in 1950, he ironically used the words 'Big Bang' to describe Lemaître's hypothesis of the primeval atom and this term, which is more accessible to the general public, has remained associated with the theory.

LEMAÎTRE'S LIFE

Photograph of Lemaître at the Catholic University of Leuven.

A KEEN INTEREST IN SCIENCE

Georges Lemaître was born to a middle-class Catholic family in Charleroi, Belgium, on 17 July 1894 and was the eldest child of Joseph Lemaître, an academic and the director of a glass and marble works, and Marguerite Lannoy, the daughter of a brewer. No aspect of his family environment set him up either for joining the clergy or for dedicating his life to mathematics and physics.

He was educated in a Christian school in the town, along with his brothers. In 1904 he began studying Classics at the Jesuit Collège du Sacré-Coeur, where he first saw the possibility of reconciling faith and science. Early on, he excelled in mathematics, physics and chemistry. When he was only nine years old, he had already decided that he wanted to dedicate his life equally to science and to God. In 1910, his family moved to Brussels and Lemaître continued his studies at the Collège Saint-Michel, where he took preparatory courses for further study. At his father's request, he passed the admissions test to study engineering and decided to leave priesthood for later.

AN INTELLECTUAL AND SPIRITUAL PATH

At the age of 17, he began studying engineering at the Catholic University of Leuven, but his studies were interrupted by the First World War (1914-1918). Shortly after the start of the war, he joined the artillery as a volunteer. He received the Belgian War Cross for his efforts during the Battle of the Yser. This challenging experience would reinforce his need to reconcile his religious and scientific vocations.

In the autumn of 1919, he returned to university and abandoned his studies in engineering to take up a new subject: physics and mathematics. In 1920, he obtained his doctorate in mathematics and his Bachelor's degree in Thomistic philosophy (i.e. philosophy inspired by the works of Thomas Aquinas). In the same year, he returned to the seminary in Malines. Fascinated by Einstein's theory of relativity, which had, however, not been widely studied in Belgium, at the same time Lemaître prepared a thesis on relativity and gravitation with the aim of winning a travel scholarship. Three years later, he was ordained a priest and received a scholarship from the Belgian government for study abroad.

FORMATIVE JOURNEYS

The young priest set off for Cambridge, England, where he studied astronomy under the famous astrophysicist Arthur Stanley Eddington, whom he greatly admired. He then crossed the Atlantic and joined Harvard College Observatory to work with Harlow Shapley (American astrophysicist, 1885-1972) on nebulae (dim, static interstellar clouds). Finally, he went to the Massachusetts Institute of Technology (MIT) to start writing a thesis on gravitational fields in fluids in general relativity.

During his studies, Lemaître was lucky enough to be part of some very active intellectual environments and to meet many key figures of contemporary physics. When he returned to Belgium in the summer of 1925, he was employed as a lecturer at the science faculty of the Catholic University of Leuven, but would continue to go frequently to England and the United States to take part in conferences or to work on his academic projects. In 1927, MIT accepted his thesis and awarded him a doctorate in physics. The University of Leuven made him a professor in the same year, a position he would hold until 1964.

ONE MAN, TWO CALLINGS

Throughout his life, Lemaître devoted himself to the Catholic faith and to science. These two callings might seem incompatible but, to him, it was entirely possible to carry out research on the beginnings of the universe without having to question his Catholic faith. While his hypothesis of the primeval atom attempted to explain the beginning of the universe's expansion, cosmology had to leave space for religion to explain the world's creation. For him, they were two truths that were independent from one another. However, his joint scientific and religious education would result in suspicion and rejection from part of the scientific community. Nonetheless, nobody could contest the quality of his research or his quest for the 'double conception' of truth, in which religion and science were separated and targeted different levels of understanding: he thus distinguished between origin, which is a physical notion, and creation, which is a philosophical concept.

An incredibly gifted mathematician, Lemaître always needed to support theory with observation. Thus, he was not happy simply constructing valid hypotheses, but verified their basis through experiments. This character trait was something that differentiated him from the other scientists of his time and allowed him to make significant advances in his research.

Lemaître died of leukaemia on 20 June 1966 in Leuven, having learned a few days earlier that the observation of cosmic background radiation had just confirmed for certain that the universe had started with an explosion, which he had already hypothesised in his theory 30 years previously.

LEMAÎTRE'S THEORIES

EINSTEIN AND FRIEDMANN'S CONTRIBUTIONS

Although Lemaître is the uncontested father of the Big Bang theory, two other scientists also played a key role in the development of this theory, which revolutionised modern cosmology. Lemaître's research, in fact, took place within a very precise scientific context.

Einstein was the first to pave the way with his theory of general relativity (1915), which was a new theory on gravitation. According to his theory, gravitational forces between objects are what gives the universe its structure. Einstein also wrote the equations governing the physical and geometric properties of the universe, which he considered to be static (meaning that the total size of the universe does not change over time). These equations enabled the way in which space changes over time to be determined, based on the amount of matter and energy that changes inside it.

Portrait of Albert Einstein in 1947.

The Russian physicist and mathematician Alexander Friedmann (1888-1925) found the solutions to these equations, which described the variation of space in time. He theorised that it was possible that the universe originated from a singularity and that, by extension, there would eventually be an end to the universe. Friedmann also estimated the age of the universe at 10 billion years. However, in the 1920s, accepted scientific estimations did not exceed one billion.

Portrait of Alexander Friedmann.

Having been exposed to Einstein's theories while studying at

Cambridge, Lemaître also became interested in the German physicist's equations and added his contribution to the new cosmology that was being established and was known as relativistic cosmology.

> **DID YOU KNOW?**
>
> Lemaître was not the only person to consider the idea of the universe's expansion. In 1922 and 1924, Friedmann published two theses in which he put forward his hypothesis of expansion. But his writings did not gain sufficient exposure. Lemaître did not gain access to these articles until 1927, at the same time that he published his own theory on the universe's expansion. We may thus say that both scientists arrived at the idea of expansion independently.

THE THEORY OF THE EXPANSION OF THE UNIVERSE (1927)

The Belgian researcher managed, independently of Friedmann, to solve the equations suggested by Einstein and found non-static cosmological solutions. He attributed a force of cosmic repulsion that forces particles in the universe to separate over time to the cosmological constant present in Einstein's equation. He also had the audacity to consider observations made by Americans at the time on the speed of nebulae, which proved that the universe truly was expanding.

Thus, in 1927, Lemaître published his key article, "A homogeneous universe of constant mass and growing radius accounting for the radial velocity of extragalactic nebulae". This uninspiring title (at least to non-experts) indicated that he had made a connection between the expansion of the universe and observations on the speed of nebulae. In the article, Lemaître described an expanding universe which, going far enough back in time, resembled Einstein's static universe. As this theory was based on observations, it was presented as the solution to Einstein's equations. Lemaître's article, however, was not as successful as it ought to have been, as Einstein himself was not convinced by his theory. It was not until 1930 that Eddington understood the significance of his former student's work. It was, moreover, thanks to him that the idea of an expanding universe became widespread.

THE HYPOTHESIS OF THE PRIMEVAL ATOM (1931)

As a continuation of this idea of an expanding universe, Lemaître put forward the idea that, in the beginning, the universe must have been much denser. As the universe is simply expanding over time, if we go far back enough, the universe must have been less and less expansive: this is what led him to reflect on the origins of the universe. According to him, the expansion of the universe must have begun from a singular initial state, that of the primeval atom. In his article "Expansion of the Universe", published in 1931, he developed this idea, again based on observations.

He believed that the very existence of nebulae meant that the universe had previously undergone processes of contraction. Thus, two opposing cosmic forces were behind the creation of the world: gravitation, which attracts, and the cosmological constant, which repels. He suggested that the evolution of the universe happens in three phases:

- The first consisted of rapid, explosive expansion following the break-up of the primeval atom.
- The second was a period of deceleration during which the density of matter and the cosmological constant became balanced. During this phase, the great structures of the universe such as the stars, galaxies and clusters formed.
- These formations are set to disrupt the equilibrium and lead to the final phase, a second rapid expansion.

This hypothesis of the primeval atom did not satisfy either Einstein or Eddington as, to them, speaking of the origins of the universe was unthinkable, since it was static. Lemaître would have to convince these leading scientists. With the help of the most recent developments in quantum mechanics, the Belgian priest decided to explain the origins of the universe through quantum theory. He concentrated on the two principles of thermodynamics (a branch of physics concerning the systems in which variations occur in the quantity of heat over time):

- energy exists in distinct quanta and the total amount of energy remains constant;
- the number of quanta is continuously increasing.

If we go back in time, we therefore find fewer quanta, which still nonetheless contain all the energy in the universe, and eventually arrive at one quantum with an extremely concentrated amount of energy, the primeval atom. Lemaître's innovative idea was to link the infinitely large (the universe) with the infinitely small (the atom).

While Lemaître's idea, which aims to explain the expansion of the universe as being due to an initial explosion, is still widely accepted, his theory that the entire universe was originally contained within a single atom which disintegrated has now been called into question. Physicists are now leaning more towards a sort of cloud of elementary particles (quarks and leptons) which gradually condensed, releasing energy and giving the universe its initial momentum. They recognise the existence of the cosmic microwave background, a trace of the initial explosion, but believe that it comes from an electromagnetic wave and not, as Lemaître thought, from a trail of particles propelled by the disintegration of the initial atom.

THE REST OF HIS WORK

After publishing his two theories, which together would form what is now commonly referred to as the Big Bang theory, Lemaître continued his cosmological research. Several of his estimations were later confirmed by scientists. He theorised on black holes and vacuum energy, as well as producing a hypothesis that the universe has additional dimensions.

After the Second World War (1939-1945), Lemaître gradually withdrew from international research by limiting his travels. He also gave up his research on cosmology for another domain he was particularly fond of and talented in: numerical analysis.

Despite the importance of his other works, he remains well-known above all for being behind this relativistic cosmology which is characterised by three main principles:

- the universe is expanding;
- the universe had a beginning;
- quantum physics (the science of the infinitely small) and astronomy (the science of the infinitely large) are linked in the understanding of the universe.

DID YOU KNOW?

Although his contribution to relativistic cosmology can no longer be contested today, Lemaître was overlooked for many years. In fact, many scientific encyclopaedias do not even mention the priest's name, or play down the impact of his work. His initial studies in mathematics and his religious commitments perhaps did not work in his favour.

THE CHRONOLOGY OF THE BIG BANG

Since Lemaître wrote his articles, scientists have revisited and corrected his conception of the Big Bang theory. The current state of knowledge on the origins of the universe is as follows.

The universe began 13.7 billion years ago in an extremely hot environment, around 1032 Kelvin (approximately 758°C). The universe was then made up only of photons, elementary particles and their antiparticles. Following an initial cosmic shock – the famous Big Bang – the particles and antiparticles disintegrated, leaving a small surplus of matter that would lead to the creation of the universe. In the first three minutes, protons and neutrons formed thanks to the presence of quarks (elementary particles). It would take 380 000 years for the universe to cool down again. This is when light was released from the matter by what scientists call cosmic microwaves. Galaxies formed following gravitational collapses of clouds of dust. Finally, stars were born, surrounded by planets.

> **DID YOU KNOW?**
>
> These cosmic microwaves can still be observed as background radiation today and constitute a vestige of the beginning of the universe. The American physicists Robert Wilson (1936-2002) and Arno Penzias (born 1933) were the first to observe this cosmic microwave background in 1965. This discovery was as fortunate as it was accidental: the physicists were actually working on creating a new type of telephone antenna. Their findings caused critics of the Big Bang theory to change their minds and support it.

After Georges Lemaître, physicists found equations that allowed them to describe the universe just 10^{-43} seconds after it began. The period that separated the hypothetical 'time zero' and 10^{-43} seconds later is called the Planck era, in homage to Max Planck (German physicist, 1858-1947), and cannot be explained by current theories, as the notions of space and time have not yet been defined. We do not know anything about that time.

Photograph of Max Planck, taken in 1933.

IMPACT

RECEPTION AMONG THE SCIENTIFIC COMMUNITY: BETWEEN CONTINUITY AND CRITICISM

The Big Bang theory appeared at a time of cosmological crisis in the scientific community regarding its representation of space. Lemaître's findings, helped by relativistic scientists, gave rise to what was practically a scientific revolution – one which would nonetheless attract some criticism – as they offered an entirely new way of conceiving the universe.

Even Lemaître's guiding authorities, Einstein and Eddington, were dubious about his theory of expansion. It took Einstein 10 years to accept the idea of an evolving universe, but he would never accept the hypothesis of the primeval atom as, in his opinion, the Belgian priest had been inspired by the Biblical story of creation, which was unacceptable for a scientific theory. In the 1940s, the theory was even discredited as it had not been confirmed by any observations. Due to a lack of proof, it would see competition from two new theories: the resurgence of Newtonian cosmology and the steady-state theory.

It took 30 years for the majority of the scientific community to recognise Lemaître's contributions to modern cosmology. It was notably thanks to George Gamow (Russian-American physicist, 1904-1968) that the Big Bang theory became well-known. He was a prolific author and keenly interested in astronomy, particularly the evolution of stars. In a text written in 1948, he developed the model of the universe in which it was dominated by heat and radiation. Gamow agreed with Lemaître's claims as to the extremely dense origins of the universe, and added that it was also extremely hot during that period. The notion of temperature allowed a key link to be made between cosmology and high-energy particle physics. He stated that all the elements in the universe were produced during the first, very hot phases of its expansion. With the help of his collaborators, Gamow calculated that in a later era in which the temperature cooled, the universe became transparent and radiation was released that can still be detected today: this is the cosmic microwave background.

Later on, notably due to the perfecting of astrophysical instruments, the following generations of scientists were able to find data that verified Lemaître and Friedmann's models. Since February 2003, the Wilkinson Microwave Anisotropy Probe has allowed the age and energy content of the universe to be calculated with great precision. Lemaître's model can therefore no longer be questioned within our current framework of knowledge.

Artist's impression of the Wilkinson Microwave Anisotropy Probe.

RECEPTION OF LEMAÎTRE'S THEORIES ACROSS THE WORLD

The 1930s were a time of various crises following the Great Depression (1929), which would lead the American media to show greater interest in cosmological discoveries, perceiving them as a way of diverting a demoralised audience. Lemaître thus became famous in 1932, when he was set in opposition to Einstein by the press. However, he would very soon be forgotten by the general public, and other scientists would wrongly be credited for Lemaître's achievements. But he was not looking for fame, and was known for his humility in the public sphere.

WHAT REMAINS OF THE BIG BANG THEORY?

The Big Bang theory as it was conceived by Lemaître is still widely accepted. It forms the very basis of our modern cosmology, our way of perceiving and understanding the universe. Thanks to him, it is now possible to pair scientific research with religious thought. The origins of the world have thus become a scientific theory independent of any religious conviction.

While nowadays the name of the theory has replaced that of its inventor, Lemaître is nonetheless still widely known in the scientific community. An asteroid (1565) has borne his name since its discovery by a Belgian astrophysicist in 1948, and the Catholic University of Leuven paid homage to him by naming an auditorium, as well as its institute of astronomy and geophysics, after him. More recently, the European Space Agency did the same for its latest Automated Transfer Vehicle, which it sent into space on 30 July 2014.

> **DID YOU KNOW?**
>
> *The Big Bang Theory*, a famous American series which started in 2007, depicts the lives of four research physicists. It is set in the United States, in Pasadena, the very town in which Lemaître met Einstein several times at the California Institute of Technology.

SUMMARY

- By establishing a new way of perceiving the universe, Albert Einstein, Alexander Friedmann and Georges Lemaître were behind a veritable scientific revolution.
- Lemaître was responsible for three ideas of new or relativistic cosmology: that the universe had a beginning, that it is continuously expanding and that quantum physics (the science of the infinitely small) and astronomy (the science of the infinitely large) are linked in the understanding of the universe.
- The theory of the expansion of the universe and the hypothesis of the primeval atom are nowadays known as the Big Bang theory.
- The origin of the universe will have had three phases: an explosive cosmic shock which enabled rapid expansion, followed by a long period of cooling down in which the universe continues to expand, followed by a second rapid expansion.
- Despite criticism, Lemaître's ideas were confirmed by the discovery of the cosmic microwave background in 1965, which had already been predicted by Gamow.

*We want to hear from you!
Leave a comment on your online library
and share your favourite books on social media!*

FURTHER READING

BIBLIOGRAPHY

- Engel, V. (2013) *Le prêtre et le Big Bang*. Paris: JC Lattès.
- Lambert, D. (2016) *The Atom of the Universe: The Life and Work of Georges Lemaître.* Krakow: Copernicus Center Press.
- Luminet, J.-P. (2004) *L'invention du Big Bang*. Paris: Seuil.
- Robredo, J.-F. (2011) *Les metamorphoses du ciel : de Giordano Bruno à l'Abbé Lemaître.*
- Université Catholique de Louvain (No date) *Georges Lemaître.* [Online]. [Accessed 3 May 2015]. Available from: <https://www.uclouvain.be/316446.html>

ADDITIONAL SOURCES

- Farrell, J. (2006) *Day Without Yesterday: Lemaître, Einstein, and the Birth of Modern Cosmology.* New York: Basic Books.
- Trasancos, S. (2016) *Particles of Faith: A Catholic Guide to Navigating Science.* Indiana: Ave Maria Press.

ICONOGRAPHIC SOURCES

- Portrait of Galileo by the painter Justus Sustermans. Royalty-free reproduction picture.
- Photograph of Lemaître at the Catholic University of Leuven. Royalty-free reproduction picture.
- Portrait of Albert Einstein in 1947. Royalty-free reproduction picture.

- Portrait of Alexander Friedmann. Royalty-free reproduction picture.
- Photograph of Max Planck, taken in 1933. Royalty-free reproduction picture.
- Artist's impression of the Wilkinson Microwave Anisotropy Probe. Royalty-free reproduction picture.

50MINUTES.com

History

Business

Coaching

IMPROVE YOUR GENERAL KNOWLEDGE
IN A BLINK OF AN EYE !

www.50minutes.com

© 50MINUTES.com, 2016. All rights reserved.

www.50minutes.com

Ebook EAN: 9782806294807

Paperback EAN: 9782806294814

Legal Deposit: D/2017/12603/130

Cover: © Primento

Digital conception by Primento, the digital partner of publishers.